T0266754

SPACE ODDITIES

First published in 2024 by the National Maritime Museum, Park Row, Greenwich, London, SE10 9NF

ISBN: 978-1-7391542-1-9

At the heart of the UNESCO World Heritage Site of Maritime Greenwich are the four world-class attractions of Royal Museums Greenwich – the National Maritime Museum, the Royal Observatory, the Queen's House and *Cutty Sark*.

rmg.co.uk

A CIP catalogue record for this book is available from the British Library.

Image credits: Front cover: astronaut (top, middle) and satellite (right, second down) ©Rawpixel; Armillary sphere (bottom right) public domain CC0 image, courtesy of Rawpixel. All other images courtesy of stock.adobe.com

Design by Matt Windsor
Typesetting by ePub KNOWHOW
Printed and bound by DZS

10 9 8 7 6 5 4 3 2 1

SPACE ODDITIES

An Astronomy Miscellany

WHERE TO BEGIN?

There is no definition of 'space' that won't start a bit of an argument, or at least a discussion! This book contains some of our favourite facts, stories and explanations of the vast Universe beyond Planet Earth, although, actually, so much of what we know about space relates to human beings interacting with it. We are part of space, as much as any distant star or hurtling comet.

For thousands of years people all over the world have considered the 'bigger picture', the cosmos surrounding us; home to a vast variety of objects and environments both tantalisingly similar to and astonishingly different from Earth. And in the modern era, humanity has journeyed into space itself to relay information and experiences earthwards.

A comprehensive book of space is probably doomed to fail; we know an enormous amount, but it is also clear that there are vast quantities of cosmic knowledge we don't yet possess. What follows, then, is a small selection of the things we find interesting. And perhaps an encouragement to find out even more...

THE SCALE OF SQUIRRELS IN SPACE

The park surrounding Royal Observatory Greenwich is full of grey squirrels. They don't seem particularly interested in learning about space,[1] but perhaps if we presented them with some facts to which they could relate, they'd be more ambitious...

Factoring in some economies of scale, it is conceivable that the per-squirrel cost of getting into space could go as low as about £600, given how light they are (big savings on fuel, compared to a human astronaut). What a bargain! And if we could get them to the International Space Station (ISS), they'd find a home roughly 180 times wider than the large drey in which they usually live. That's about 360 times their body length – positively palatial.

[1] South American squirrel monkeys have been into space, but that doesn't count.

SQUIRREL FACT FILE (APPROXIMATE):

Length: 30cm (12 in.), not counting the tail!
Weight: 0.5kg (1lb 2oz)
Speed: 32km/h (20mph)
Life span: 9 years
Habitat: Grey squirrel dreys are about 60cm (24 in.) wide

Unfortunately, there is no tree branch between Earth and the Moon, but if there was it would take a squirrel about 17 months to run along it at top speed. When it got to the Moon, it would feel as if it weighed about the same as four AA batteries here on Earth.

It wouldn't live long enough to see Jupiter orbit the Sun even once, sadly, but it would see a little more than 100 Full Moons.

Let's go bigger:

The Milky Way is over 3 sextillion (3,000,000,000,000,000,000,000) squirrels wide.

Light travels roughly 35 million times faster than a squirrel.

The Sun has the same mass as about 4 nonillion (4,000,000,000,000,000,000,000,000,000,000) squirrels.

The Big Bang happened approximately 1.5 billion squirrel lifetimes ago (though there weren't any squirrels around then, of course...).

THE PERIODIC TABLE

The Universe is a complex place, composed of a dizzying array of structures and objects themselves made from a little less than 100 naturally occurring elements – and that's if we limit ourselves only to things composed of normal matter. Keeping track of all of this isn't exactly an easy task, but there are tools to help.

The periodic table was an attempt to bring order to what seemed a very complicated system. By recognising patterns in the properties of certain groups of chemical elements, a number of scientists, most notably Dmitri Mendeleev in the nineteenth century, put sets of similar elements together. The table was initially

incomplete, containing gaps suggesting where additional elements, which hadn't yet been found, could be. Over time, the predictions mostly came true and the table we have today contains 118 separate elements.

It is extremely useful and possessed of its own kind of beauty... but it is quite complicated. And while it wouldn't be fair to claim that all astronomers are lazy, they do often try to simplify issues where possible. By mass, some 74 per cent of the normal matter in the Universe is hydrogen while 24 per cent of it is helium,

leaving a paltry 2 per cent for everything else! And so, to an astronomer, it is often useful to classify the make-up of much of the Universe simply

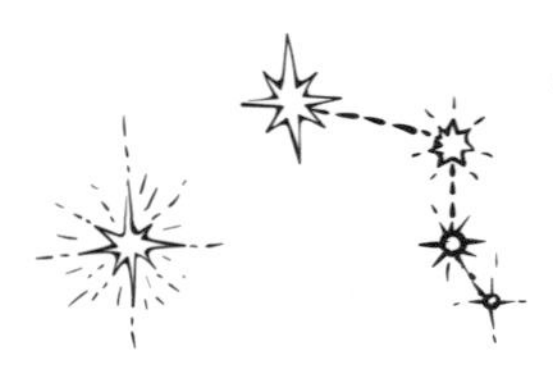

as hydrogen, helium and metals – a convenient term for all the rest.

But what is it that has made the Universe so lopsided in its composition? Well, different conditions are required to form each element. Smaller, simpler elements are usually easier to produce, requiring lower temperatures and pressures in whatever cosmic furnace they originate in. The periodic table, then, is also a rough map of where the elements in the Universe come from.

Most of an atom's mass comes from the subatomic particles that make it up – from the protons and neutrons in its nucleus, with a small contribution from electrons. Comprised

of a single proton (with or without an electron), hydrogen is the lightest element, the first entry in the periodic table. As soon as protons could form – a small fraction of a second after the Big Bang – hydrogen, technically, existed.

Helium, slightly heavier, also requires relatively little assembly, though things do need to be hot and dense enough for it to take place. Within the first ten minutes after the Big Bang, the Universe was hot enough to fuse together hydrogen to make helium, but not so hot that the new element would instantly be torn apart again. The merest hint of lithium was also produced during this time, but, otherwise, in the time it takes to overboil an egg, the Universe had formed around 98 per cent of the atoms that would ever be formed.

Many elements are created in the cores of stars, the immense heat and pressure allowing lighter elements to fuse together into atomically heavier ones. These 'element factories' can produce layers of increasingly heavy elements until they are creating iron in their cores. The slowly increasing temperatures that are found as these behemoths evolve towards their eventual explosive ends make the perfect pressure cooker.

Beyond iron, though, conditions inside these stars aren't sufficient for the formation of heavier elements. Fusing iron actually requires more energy than it produces, so stars that make iron in their cores rapidly cool, collapse and explode in a supernova. This huge burst

NEUTRON STARS

Neutron stars are the smallest and densest stars in the Universe. They are the remains of massive stars whose cores have collapsed and been crushed into objects only around 20km (13 miles) across. We find them scattered everywhere in our Galaxy and some appear to pulse. These rotating neutron stars give off jets of radiation and are known as pulsars.

of energy allows heavier elements to form while also topping up the quantities of lighter elements already made. Elements like arsenic and gallium are almost entirely produced through these cosmic explosions.

Even heavier elements are created when two neutron stars collide, causing a new explosion known as a short gamma-ray burst and producing lots of neutron-rich elements.

ANTIMATTER

Our Universe is filled with matter, composed of particles such as protons and electrons, which have qualities like mass and charge. Antimatter is comprised of particles almost identical to their matter counterparts, but with some opposite characteristics. Every particle has its own antiparticle, and we have observed them through various experiments, but antimatter doesn't last long. When antimatter and matter are close enough, they collide and disappear – their mass converted into energy through a process called annihilation. Considering that every particle has its 'anti' partner, the Universe should be filled with energy made from these annihilation processes. However, our Universe is dominated by matter and scientists are still trying to understand why there is such a shortage of antimatter.

Uranium, platinum and gold are all formed in these rare encounters.

Finally, the heaviest elements in the periodic table are not naturally occurring, existing only for mere moments in the extreme environments within human-constructed particle accelerators before breaking apart to create something more stable again. These 'synthetic elements' include Americium and Tennessine.

But all the rest, some 94 or 98 elements (depends on who you ask), exists at least in traces throughout the Universe, produced by a wide range of different processes, almost all of which require the death of at least one star. Lovely.

THIS *IS* ROCKET SCIENCE!

Rocket science isn't as tricky as its reputation suggests. Sure, getting all the details right to successfully launch a crewed spacecraft into the exact orbit needed for a chosen mission involves a lot of complex maths and engineering, but the general principles of getting something into space are actually pretty simple.

If we want to travel up from Earth's surface, we have to overcome what's holding us down – gravity! To do this, we need to produce a greater amount of force and use it to push our rocket upwards.

This is where some laws of physics come in, specifically Isaac Newton's Third Law of Motion. In it, Newton said that every force has an equal but opposite reaction; if you produce a force in one direction, you will move the

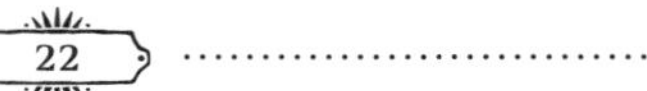

opposite way with the same power that you provided. A good way to picture this is to think about rowing. When you row, you push the water backwards with your oars and it moves the boat forwards. We do the same thing when we launch things into space – to push our rocket up, we create a downwards force.

Most rockets use chemistry to do this. The chosen fuel is activated when it's time to go, usually by combining it with oxygen to start a reaction. This process also creates heat, which increases the pressure inside the rocket and forces out the products of the reaction at one end.

We've created our force! We've overcome gravity! Our rocket can make it up into space, dropping fuel tanks back to Earth once they're empty.

Now to get back down to the surface again... Luckily, gravity is going to help us this time. Unluckily, the speed gravity wants us to move at isn't particularly safe.

To slow our re-entry capsule down, we can use Newton's Third Law again. Gravity is still pulling us down, so we want to produce some downwards force to act on our spacecraft in the upwards direction. We can use mini boosters or retrorockets to do this as they work in the same way as our main launch rocket, just with less fuel. We don't need as much fuel this time as we don't want to completely overcome gravity and travel back up again! Early space missions

WHERE WE'RE GOING, WE DON'T NEED ROADS

That's certainly true for the Tesla Roadster that was launched into space on 6 February 2018. Selected as a dummy payload (in place of an expensive satellite, or even more valuable astronauts...) for the first SpaceX Falcon Heavy *test flight, the Roadster survived the launch and, along with a mannequin nicknamed 'Starman', was sent cruising into orbit around the Sun.*

favoured parachutes to aid the final parts of the descent stages (and often an ocean splashdown to help further), but modern missions are being designed to include safe, controlled return sequences entirely reliant on the principles of rocketry.

THERE'S A ROCK ON MARS CALLED HUMPHREY

The rules for naming planetary features are organised by the International Astronomical Union Working Group for Planetary System Nomenclature (or the IAU WGPSN – clearly the working group on naming the working group didn't go so well). Different themes are adhered to for different types and sizes of feature.

CRATERS AND VALLEYS

Large craters on Mars should be named after scientists who have 'contributed significantly to the study of Mars', or 'writers and others who have contributed to the lore of Mars'. For example, the *Curiosity* rover is currently exploring Gale Crater, named for Walter Frederick Gale (1865–1945), an Australian astronomer, and there is another feature called

Bradbury Crater, named after Ray Bradbury (1920–2012), the American science fiction author whose work includes *The Martian Chronicles.*

Large valleys, which are found on Mars due to the flow of liquid water across the planet in the distant past, should be named after the word for Mars in different languages. Hence, we have Ares Vallis, Marikh Vallis and Ma'adim Vallis, all meaning 'Mars Valley' in Greek, Malay and Hebrew respectively.

MARTIAN MOONS

Features on Phobos, the largest of Mars's two moons, are named either after scientists who studied the Martian satellites, or people and places from the 1726 novel *Gulliver's Travels* by Jonathan Swift (1667–1745). Among other fantastical and

satirical adventures, Swift's novel features fictional scientists on an island called Laputa who have discovered two moons of Mars, though the orbital characteristics aren't actually correct and the fact that he even chose two moons was probably luck, as Mars's moons weren't discovered in real life until 151 years after the story's publication! There is a crater on Phobos called Gulliver, after Lemuel Gulliver, the main character in the book.

There is another called Stickney, after American mathematician and suffragist Angeline Stickney (1830–92). She was also the wife of Asaph Hall, who was credited with the discovery of Phobos and Deimos, the second Martian satellite, in 1877. Angeline gave up her academic career when she married Asaph, as

was standard at the time, but helped him with mathematical calculations. Asaph claimed he would have given up the search for the moons without Angeline's support but refused to pay her a salary for her computational work as she requested. After this she declined to support his research any further.

ROVER DISCOVERIES

Depending on your point of view, you might think this themed system pleasing or limiting, but for a glimpse of what naming would be like without the oversight of the IAU we can consider the names of individual rocks

explored by the rovers currently on the Red Planet. The IAU doesn't get involved with rocks less than 100m (328ft) in size, so names are left to the scientists and engineers studying them. This has led to unofficial names such as Home Plate, because the rock was shaped roughly like the bases in a baseball game; Bounce, as one of the airbags used for the safe descent of the *Opportunity* rover bounced on, or near,

ASHES TO ASHES #1

When NASA's New Horizons *spacecraft began its almost decade-long journey to the dwarf planet Pluto in 2006, it carried a very special payload along with it – some of the cremated remains of Clyde Tombaugh (1906–97) contained within a small canister. It was only fitting that the astronomer who discovered Pluto back in 1930 was part of the mission that brought us humanity's first, and (so far) only, close-up views of this distant world.*

this rock and Jake M, which was named after a *Curiosity* rover engineer who sadly died just days after the rover landed. There are others called Oileán Ruaidh, after an island off the coast of Ireland which translates to 'Red Island'; Hottah, named for Hottah Lake in Canada and Humphrey, after Humphreys Peak, the tallest mountain in Arizona in the United States. There are even Martian rocks named Big Joe – presumably quite a big rock – and Barnacle Bill (no idea...).

There are no universal naming conventions, particularly because new discoveries keep being made. The best we can hope for is some semblance of coherence! You'll find some other examples later on.

THE ELECTROMAGNETIC SPECTRUM

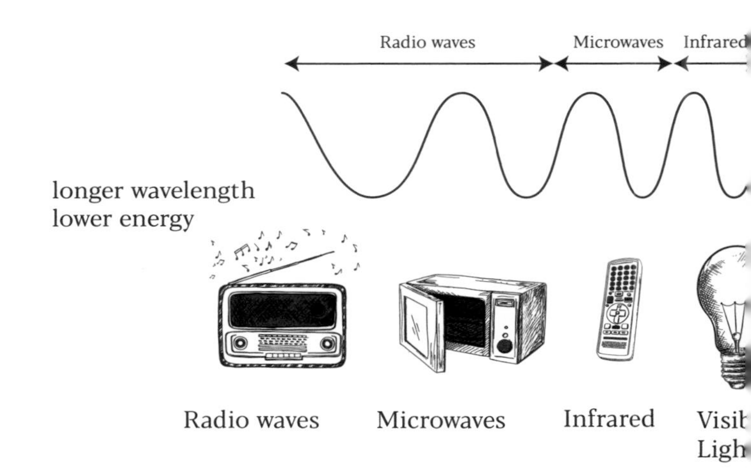

Human beings have evolved to see what is known as 'visible light', but there are other types of light with varying colours, wavelengths and levels of energy that we are blind to. The full range of wavelengths, including visible light, is known as the electromagnetic spectrum. Different physical processes produce radiation at different wavelengths, so to get as complete a picture of the Universe as possible, astronomers use different types of telescopes to observe astronomical

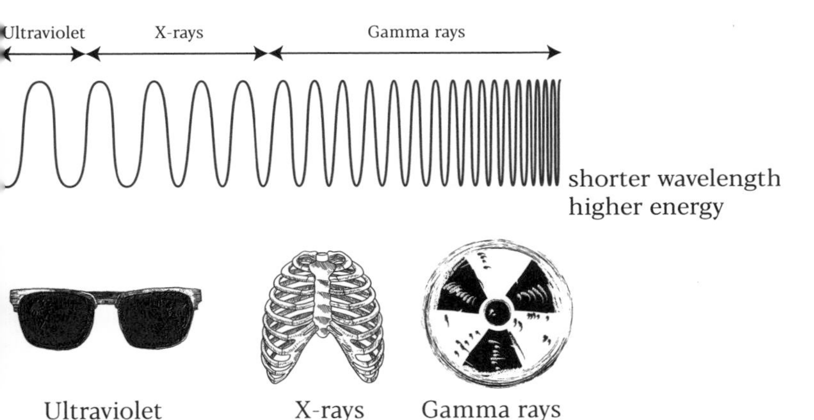

objects in as many different parts of the spectrum as they can.

GAMMA RAYS

At the extreme short-wavelength end of the spectrum are gamma rays, emitted by the most energetic objects and events in space. Some of the most mysterious are gamma-ray bursts (GRBs) – the brightest and most violent explosions human beings have ever observed in the Universe. They typically last from a fraction

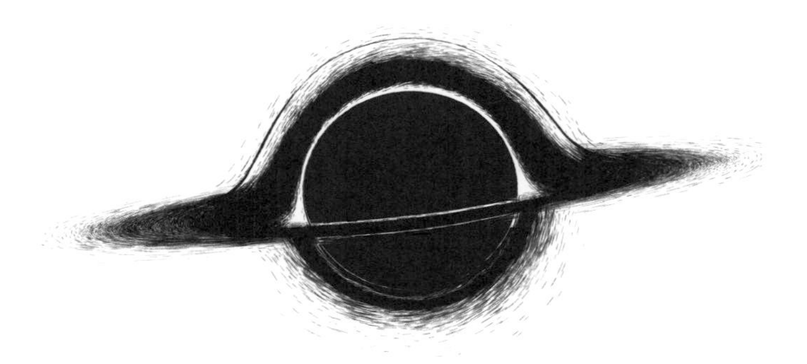

BLACK HOLES

Black holes are the final products of the deaths of massive stars, areas of space with such intense gravity that nothing, not even light, can escape. They are difficult to spot, as they do not produce any light, but we can see the effects of them. The light from distant stars will bend around black holes in front of them, and matter that strays too close to a black hole will often emit radiation as it falls in. So, by observing these processes we are able to map out the region surrounding a black hole.

of a second to several minutes, but the energy released in that short time is comparable to the total emission of the Sun over its entire 10-billion-year lifespan. Shorter-lived GRBs are the tell-tale signs of the birth of a black hole after two neutron stars collide, or the result of a black hole devouring a neutron star.

X-RAYS

Like gamma rays, x-rays are emitted by very hot and energetic objects and events. Many of these are found far away from our Solar System but our cosmic neighbourhood also has its fair share of x-ray sources. If we had x-ray vision and looked at Jupiter, we would see its entire face glow in x-rays as the planet's hydrogen-rich atmosphere scatters them from the Sun.

ULTRAVIOLET (UV)

Our Sun, like every star in the Universe, emits ultraviolet – or UV – light. It's the Sun's ultraviolet light that leads to sunburn for those who are exposed to strong sunlight for too long. Fortunately, Earth's atmosphere blocks the majority of the Sun's ultraviolet output, protecting us from harm. Young stars emit more ultraviolet light than older ones, so astronomers can map where young stars are distributed in a galaxy by looking for regions that are bright in this part of the spectrum.

VISIBLE LIGHT

Visible light has been used in astronomy since ancient times. In fact, other wavelengths in the electromagnetic spectrum were only discovered during the nineteenth and early twentieth

centuries. We have given a vast number of names to the different parts of this range of wavelength and refer to them as colours! Stars have different colours ranging from red, orange and yellow to blue and white. These colours allow astronomers to measure the temperature of a star's surface without the need for a thermometer. Blue supergiant stars like Rigel have surface temperatures of around 12,000°C (21,600°F), while our yellow Sun's temperature is cooler at about 5,500°C (9,930°F) and Betelgeuse, the red supergiant star, is comparatively frigid with a surface temperature of 3,000°C (5,430°F).

INFRARED

Space is surprisingly dusty. When astronomers only studied the cosmos using

visible light, dust clouds found in nebulae shrouded the formative periods of stars and distant planets in mystery. Infrared light has longer wavelengths than visible light, which means it is not as easily stopped by cosmic dust. An infrared telescope can detect the warm glow from baby stars hiding within a nebula.

MICROWAVES

The most well-known example of studying the Universe in microwave wavelengths is the Cosmic Microwave Background (CMB). Over the last 14 billion years or so, this light has stretched and shifted towards the microwave part of the electromagnetic spectrum. The CMB is surprisingly

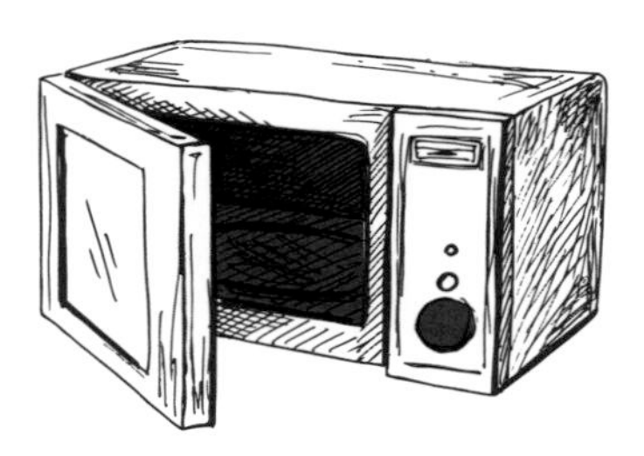

almost completely uniform throughout the Universe, although there are tiny differences in temperature (in the order of a 1,000th of a degree!) which relate to very small fluctuations in density. Regions that are cooler are denser, while those with a higher temperature have less matter. We now see galaxies and galaxy clusters in high-density regions and voids in areas with low density.

COSMIC MICROWAVE BACKGROUND

The Cosmic Microwave Background, or the CMB, is one of the key pieces of evidence we have for the Big Bang theory. It is the leftover energy from the formation of the Universe – a relic of the first light to have freely travelled through space as the radiation from the initial explosion spread out with the expansion of the Universe.

RADIO WAVES

Radio waves inhabit the least energetic, longest-wavelength part of the electromagnetic spectrum. They are emitted when fast-moving electrons interact with magnetic fields, such as those present in the Sun, Jupiter and pulsars. The first extraterrestrial radio source discovered came from the heart of the Milky Way and was later found to be a supermassive black hole called Sagittarius A*. In 2022, the Event Horizon Telescope Collaboration used radio waves to produce the first image of it.

TUDOR MOON

Space is a hostile environment, so a spacesuit is an absolute necessity for any human venturing out to explore it. You can think of a spacesuit as a one-person pressurised spacecraft equipped with vital life-support systems. In the early 1960s it became clear that if astronauts were going to set foot on the Moon, they would need far more sophisticated spacesuits than the modified pressurised flight suits in use at the time.

In 1962, the AiResearch Manufacturing Company was working on a proposal for the design and construction of a spacesuit. It needed to provide complete protection from the hazards of space, be rigid and durable, and

allow the astronaut to move freely. Quite a task, but the company soon realised that the design problems they were facing had in fact been solved centuries before.

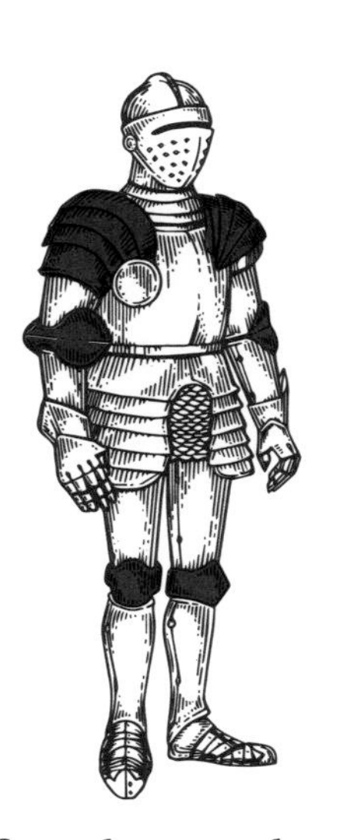

In the sixteenth century, Henry VIII commissioned a suit of armour for his Field of the Cloth of Gold meeting with Francis I of France. The royal armourers at Greenwich faced a tough challenge – the armour had to be suitable for combat on foot, providing coverage from head to toe but not restricting movement. Sound familiar? Jerry Glacer and his team at AiResearch certainly thought so.

The combat armour was a remarkable feat of engineering. By using a series of overlapping steel plates held together by sliding rivets,

the Greenwich armourers produced a fully articulated suit that provided complete protection to its wearer. Sadly, a change in tournament rules meant that Henry VIII never had a chance to wear the armour, so it was sent to the Tower of London to be put on display.

In March 1962, Glacer sent a letter to Sir James Mann, the Master of the Armouries, requesting photographs and information about the suit. Glacer and his team used the information to develop a spacesuit that they presented to NASA. Although the company's proposal was unsuccessful, in recognition of the role the suit of armour played in inspiring early spacesuit designs, in the 1970s NASA sent a spacesuit to the Royal Armouries, which was displayed next to Henry VIII's foot combat armour.

STUFF... THAT ISN'T

Astronomers love to be precise, and correct! But sometimes they get things wrong, or change their minds, or discover a new piece of information. It's all part of the adventure. Here are a few of the best, or maybe worst, examples...

VULCAN ISN'T A PLANET – OR ANYTHING, ACTUALLY

In the nineteenth century, astronomers were a little troubled that Mercury didn't behave exactly as predicted by Isaac Newton's Law of Gravitation (though it was really close). They knew that as the planets reach the point of closest approach to the Sun in their orbit (the perihelion), they 'precess', or shift slightly each time so they don't trace out exactly the

same path. This is mainly because of little gravitational pulls by other planets, but Mercury didn't move as the astronomers expected. This behaviour was particularly troublesome for Urbain le Verrier (1811–77), who had previously achieved great success by predicting the position of Neptune using only mathematics (whereafter it was seen very close to this predicted position by other astronomers – good job, Urbain!). Using historical observations of Mercury's transit across the Sun's disc, he estimated the error in Mercury's orbit to be only a small fraction of a degree per century. But not zero! He thought Mercury's motion could be explained by the presence of another planet orbiting the Sun even more closely than Mercury and providing additional gravitational influences not previously accounted for. Its proximity to the Sun, and presumed small size, would make observing the object very difficult.

ISAAC NEWTON

Attempts to discover 'Vulcan', as the object became known, went on for decades, with claims and counterclaims of discovery, changing notions about the planet's nature (was it even a planet?) and physical characteristics, and lots of dissatisfaction about the lack of repeatable, demonstrable observations.

And eventually... it turned out that using the wrong tool had created a puzzle that wasn't really there at all. Newton's Law of Gravitation couldn't account for all the factors that affect

the planets (and everything else) in real life. It was very good in lots of circumstances, which is why we still learn about it today, but it wasn't perfect! Mercury (and indeed all other planets) would not behave as it predicted.

FROM NEWTON TO EINSTEIN

Newton's formulation of gravity, his laws of motion and other theories function well as long as the objects examined aren't moving very fast or are either extremely massive or extremely small. We use them in a lot of circumstances to this day. But when considering extreme circumstances, Einstein's framework of understanding space and time as parts of an encompassing and distortable continuum is essential. Newton's ideas couldn't account for extremely strong gravity, velocities close to the speed of light or quantum mechanics, and investigating these are vital to understanding the Universe.

Albert Einstein's General Theory of Relativity, published in 1915, considered gravity in a different manner and was able to predict the observable motion of Mercury (which played a big part in Relativity's acceptance by the scientific community) without the need for adding in an 'extra' planet. Vulcan never existed!

ANDROMEDA ISN'T A NEBULA

Andromeda is visible with the naked eye under the right conditions, appearing as a little smudge in the night sky. For a long time, it was thought of as a nebula (from the Latin, meaning 'mist') – a large cloud of dust and gas. But even as astronomers trained telescopes on the object, debates raged for generations about its true nature: was it a nebula within the Milky Way? Was it something like the Milky Way? (It took a long time to understand even roughly what the Milky Way looks like, since we're inside it.) Or was it something entirely different?

The debate about the Great Andromeda Nebula, as it was being called, was finally settled in 1922 by Edwin Hubble when he identified Cepheid variable stars in Andromeda. There is a direct relationship between the luminosity of these stars and the period of time over which they dim and brighten repeatedly. In

short, if you know how long they take to vary in brightness and how bright they should be, you can quickly work out how far away they must be just by observing how bright they are. Applying this technique to the stars found in Andromeda indicated that they were at enormous distances from Earth, far beyond the Milky Way itself.

SPECTRA

The elements making up an object can absorb and emit particular wavelengths across the electromagnetic spectrum, producing characteristic patterns in the intensity of light at different wavelengths. We call these patterns spectra. By looking very closely at the spectra of light from something, you can identify what it must be made of – a vital tool for astronomers, who cannot go and touch most astronomical objects, or create their own versions in a laboratory!

It turns out that we were looking at an entirely separate galaxy, not a nebula. The Andromeda Galaxy, in fact, is about two and a half million light years away and contains around a trillion stars (more than twice as many as the Milky Way).

Pretty impressive for a little smudge...

NEBULIUM WASN'T A NEW DISCOVERY AFTER ALL

In the nineteenth century, spectra of the light from astronomical objects confused many astronomers. Some objects had spectra a bit like the Sun, while others were quite different. Observations of the Cat's Eye Nebula (also known as NGC 6543) showed light emitted at

wavelengths that didn't correspond to any known elements on Earth. Perhaps it was something completely new! And because it was seen in a nebula, some astronomers thought, 'Perhaps we could call it "nebulium"?'

Actually, nebulium 'vanished' in 1927 thanks to the work of astronomer Ira Sprague Bowen (1898–1973), because it turned out to be doubly ionised oxygen. This may not sound too exciting at first (a new element seems more fun, potentially), but outside a laboratory doubly ionised oxygen requires the extremely diffuse, near-vacuum environments of nebulae (explaining why it isn't really seen on Earth, or in the spectra of the Sun or other stars). It also fit into the – at that time, relatively new – predictions of the periodic table.

Nebulae turned out to be just a little less mysterious, but more interesting than we had thought.

THERE'S NO MOON CALLED BOTTOM

The name Uranus has been making small children giggle, presumably, for hundreds of years.

The planet was discovered in 1781 by William Herschel, who worked alongside his sister Caroline. William suggested 'Georgium Sidus' as its name, in honour of the King of England at the time, George III. 'Uranus' was proposed in 1782 by Johann Bode, another astronomer who had been studying the planet further. Even before the rules of the International Astronomical Union came into being in the early twentieth century, astronomers were really into themes so Bode argued that 'Georgium' didn't follow the pattern of the other planet names and that Uranus, after the Greek god of the sky, was a much better fit.

Herschel also discovered two moons orbiting Uranus shortly afterwards. They were named Oberon and Titania, after the king and queen of the faeries in William Shakespeare's play *A Midsummer Night's Dream*. It's unclear why they received these names, or who named them – it could have been Herschel or possibly his son John, a prominent astronomer in his own right.

Two further moons were discovered in 1851 and named Ariel and Umbriel. Umbriel is a character described as a sprite in a poem

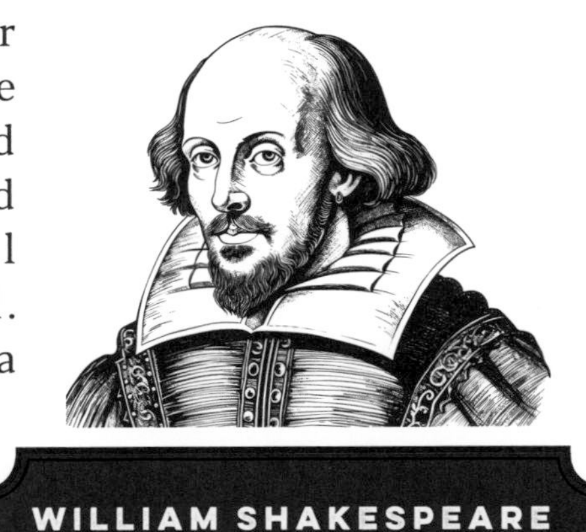

WILLIAM SHAKESPEARE

by Alexander Pope and Ariel is another spirit in the same poem, as well as a spirit in Shakespeare's play *The Tempest*.

By this point a theme was firmly established. Gerard Kuiper found another moon in 1948 and named it Miranda (heroine of *The Tempest*) and when the *Voyager 2* probe found ten new Uranian moons in 1986 astronomers stuck to the theme and used nine names of Shakespeare characters, plus one more from Pope, Belinda.

A further 13 moons have been found since then, and the naming convention has been followed. Sadly, no one has yet been brave enough to suggest Bottom, the name of another character from *A Midsummer Night's Dream* which would fit the rules perfectly.

THE INVISIBLE MILKY WAY

Our view of the Milky Way from Earth – that band of light across the night sky – is a very limited picture of our Galaxy. Not only is our view edge-on and from within, but we have to combine the data-collecting of many different telescopes to build up a reasonable store of information about what we can see.

However, there is yet another part of the Milky Way that can never be seen with any telescope, because the particles that make it up allow light to pass through it. These particles are called dark

matter, and they are known to us because of the influence they exert on the parts of the Galaxy that we can see.

The fact that we may never be able to literally shed light on this dark component of the Milky Way doesn't mean we can't try to study it and use that information to unveil the Universe around us.

WHAT TELLS US THERE'S DARK MATTER IN THE MILKY WAY?

In the 1960s and 70s, Vera Rubin, a pioneering astronomer, was studying the rotation rates of spiral galaxies by comparing the velocities of stars near

their centres with those closer to the edges. It had been thought that stellar velocities would increase rapidly from the centre of the galaxy but then decrease towards the edges.

Rubin's work showed that this wasn't the case. Stellar velocities remained high in the outer regions, suggesting an undetected gravitational influence on the outer stars that enabled them to move faster. Something was causing the effect; invisible but influential. That something is theorised to be dark matter, without which spiral galaxies would – according to Rubin's predictions – fly apart. Her calculations showed that galaxies would require between five and ten times more dark matter than ordinary matter to stop that happening. That is to say, there is far more out there than we can actually see directly, with dark matter enveloping us in an invisible halo.

DARK ENERGY

If you thought dark matter was complicated, have you heard of dark energy? The Universe is expanding and the rate of its expansion is also accelerating. We know this because galaxies are moving away from us, and the further away galaxies are, the faster they move. Astrophysicists predict that dark energy is the mechanism responsible and believe that it makes up around 68 per cent of the mass-energy content of the Universe. Dark energy remains a mystery, but a very important one, as it has such a profound effect on the Universe. It is one of the greatest challenges of modern physics.

ANIMALS IN SPACE

As monumental as Yuri Gagarin's first spaceflight was in 1961 and as iconic as Neil Armstrong's 'one small step' became in 1969, it is worth noting that human beings were not the first species to 'boldly go'. Before we dared to venture beyond the safety bubble of our planet's atmosphere, it was Earth's own humble creatures (great and small) that were tasked with testing the safety of such an endeavour. Many of these missions were motivated by the Space Race, as the United States and the Soviet Union competed for spaceflight superiority in the 1950s and 60s.

Here are just a few firsts accomplished by Earth's brave beasts and courageous critters.

FIRST LIVING CREATURES IN SPACE

As it turns out, the first Earth-bound creatures to hitch a ride to space (not including bacteria) were actually experienced pilots (of sorts) – a

swarm of fruit flies. On 20 February 1947, the insect astronauts were launched into space on board a V-2 rocket for a suborbital flight. The aim of the US-led mission was to test whether radiation from space could have a harmful impact on human astronauts. The flies returned to Earth via parachute after a flight of just over three minutes, apparently unharmed by any radiation they might have encountered, though presumably quite confused.

FIRST LIVING CREATURE TO ORBIT EARTH

Possibly the most famous animal ever to take a trip to space was Laika, a stray dog found on the streets of Moscow, who became the first animal to orbit Earth successfully. It was stated at the time

that since Laika was a stray she would be well suited to deal with the harsh and unpleasant conditions that often come with space travel, and various pre-flight tests found that she was indeed resilient.

On 3 November 1957, Laika was launched into space on board her spacecraft *Sputnik 2*. Despite completing four successful orbits of our planet, Laika sadly died just hours into the flight after an air-conditioning malfunction caused the cabin to overheat. Since her historic mission, Laika has become known internationally and her legacy lives on to this day. In 2008, a monument was built in her honour outside Star City, Russia, the cosmonaut training facility where her mission preparation took place.

FIRST LIVING CREATURES TO VISIT THE MOON

And finally, here's a true story that could perhaps serve as a bizarre sequel to Aesop's 'The Tortoise and the Hare'. On 14 September 1968, just three months prior to NASA's historic Apollo 8 mission that saw three astronauts orbit the Moon for the first time, two slow and steady reptiles set off to beat them to it. In a bid to test the effects of cosmic radiation on terrestrial lifeforms, the Soviet Union sent two tortoises on a

six-and-a-half-day return trip to the Moon on board the spacecraft *Zond 5*. The biological payload for the mission also included a variety of soil and seed samples, fruit fly eggs, worms and even a few flowers.

Thankfully, this rather whimsical team of space travellers returned to Earth safely, splashing down in the Indian Ocean on 21 September. Despite losing around 10 per cent of their body weight during the mission, the tortoises were

PLAY WELL #1

While on board the International Space Station in 2012, Japan Aerospace Exploration Agency (JAXA) astronaut Satoshi Furukawa built a model of the space station entirely out of Lego. To avoid any pieces floating away, a Lego enthusiast's worst nightmare, the model was pieced together inside a glovebox.

reported to have lost none of their appetite and enjoyed a well-deserved meal on their return.

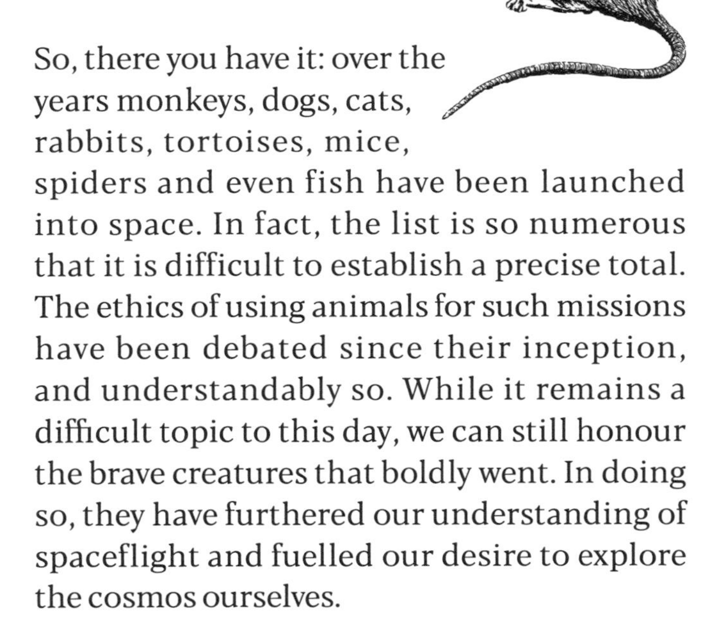

So, there you have it: over the years monkeys, dogs, cats, rabbits, tortoises, mice, spiders and even fish have been launched into space. In fact, the list is so numerous that it is difficult to establish a precise total. The ethics of using animals for such missions have been debated since their inception, and understandably so. While it remains a difficult topic to this day, we can still honour the brave creatures that boldly went. In doing so, they have furthered our understanding of spaceflight and fuelled our desire to explore the cosmos ourselves.

EXTREME EXOPLANETS

Exoplanets are planets beyond our Solar System. Although the first confirmed detection wasn't made until 1992, we have now detected thousands, and are adding to the list all the time. Whenever we write something about exoplanets, another comes along and 'beats' it pretty quickly. So here are some facts about a few exoplanets compared to the planets in the Solar System. By the time you finish reading this, there may already be even more extreme examples!

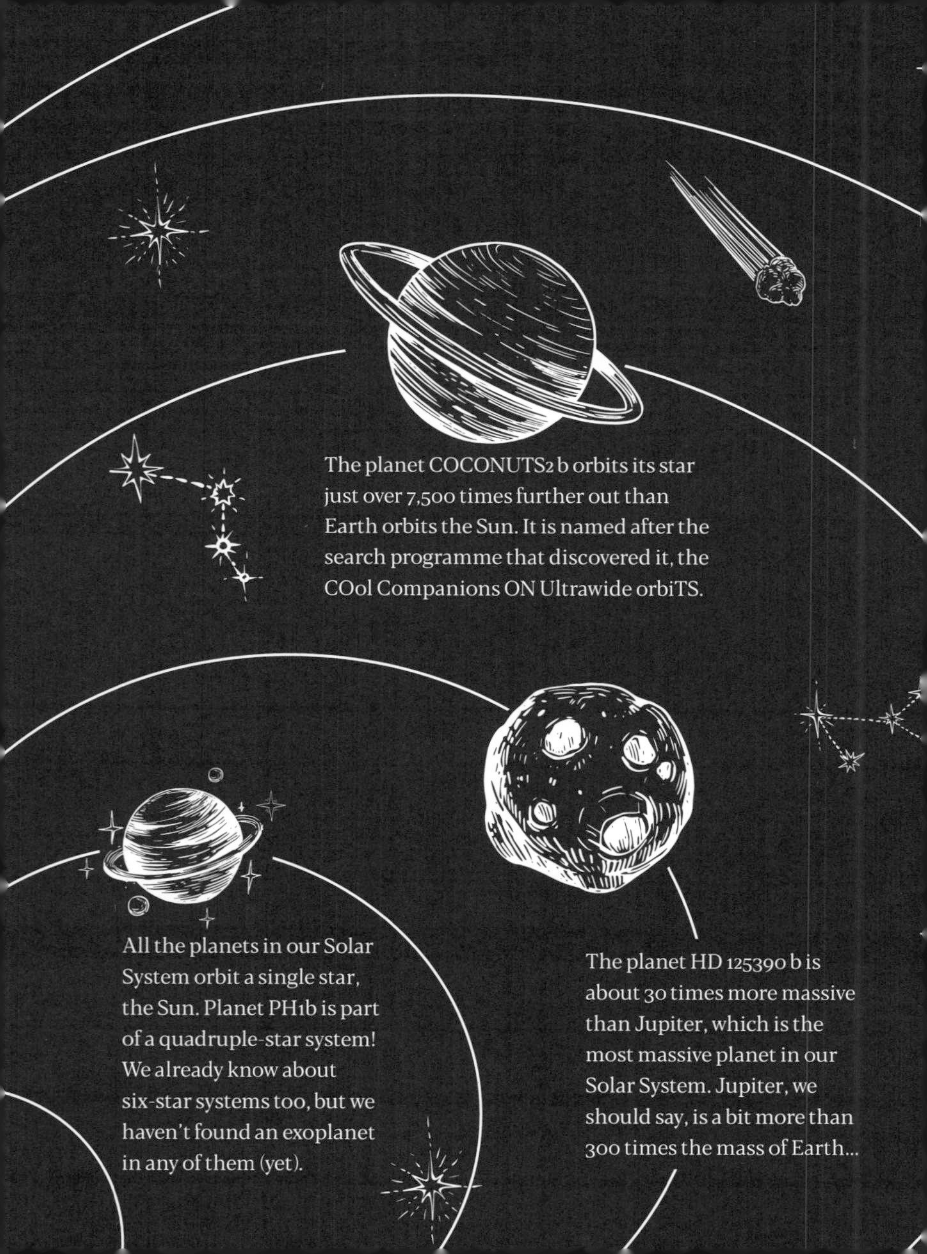

The planet COCONUTS2 b orbits its star just over 7,500 times further out than Earth orbits the Sun. It is named after the search programme that discovered it, the COol Companions ON Ultrawide orbiTS.

All the planets in our Solar System orbit a single star, the Sun. Planet PH1b is part of a quadruple-star system! We already know about six-star systems too, but we haven't found an exoplanet in any of them (yet).

The planet HD 125390 b is about 30 times more massive than Jupiter, which is the most massive planet in our Solar System. Jupiter, we should say, is a bit more than 300 times the mass of Earth...

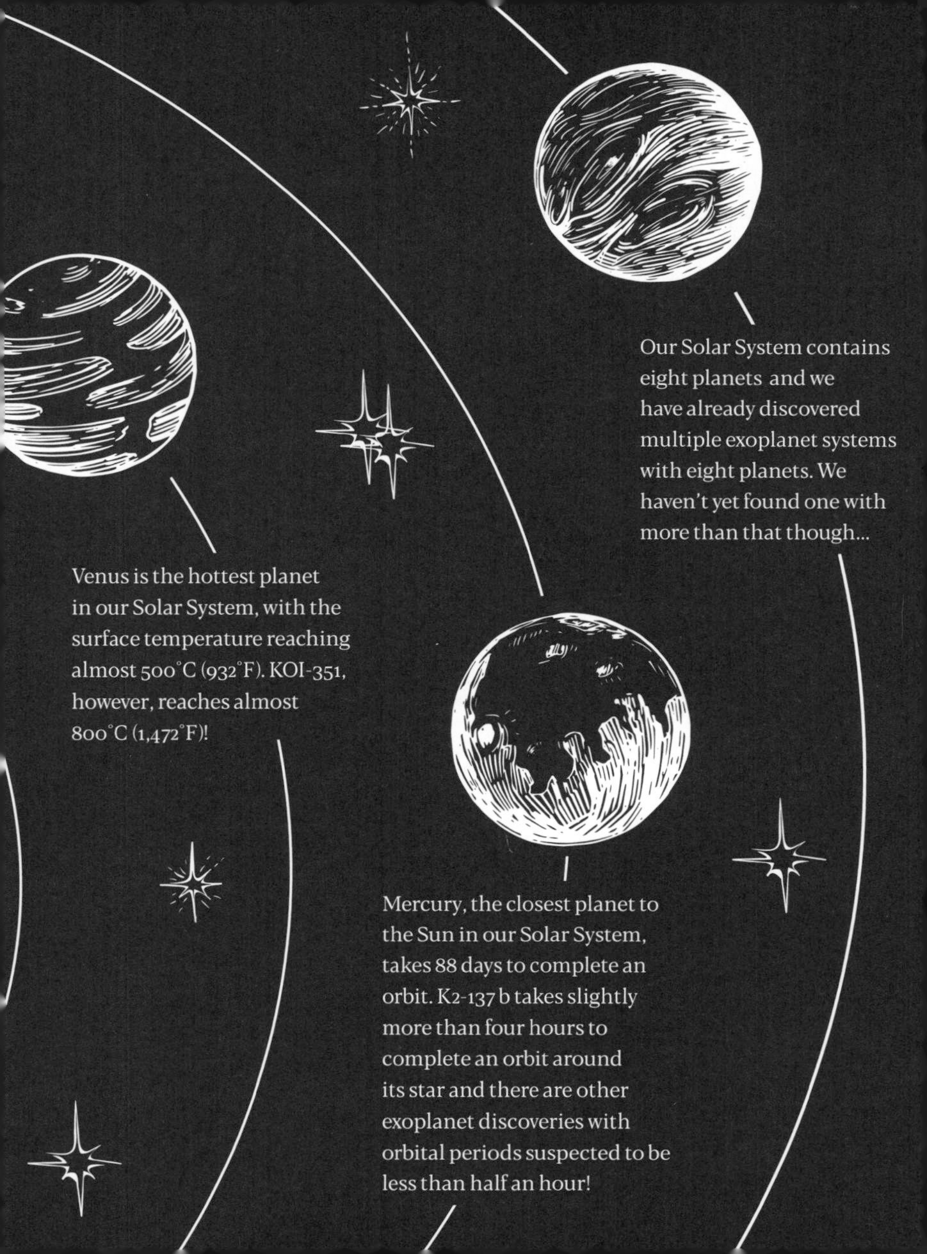

Our Solar System contains eight planets and we have already discovered multiple exoplanet systems with eight planets. We haven't yet found one with more than that though...

Venus is the hottest planet in our Solar System, with the surface temperature reaching almost 500°C (932°F). KOI-351, however, reaches almost 800°C (1,472°F)!

Mercury, the closest planet to the Sun in our Solar System, takes 88 days to complete an orbit. K2-137 b takes slightly more than four hours to complete an orbit around its star and there are other exoplanet discoveries with orbital periods suspected to be less than half an hour!

THE COSMIC WEB

Galaxies are vast structures that contain gas, dust and stars. They vary in shape and size and serve as homes to billions of star systems throughout our Universe. It wasn't until the 1920s that astronomers realised some of the 'nebulae' they had catalogued were in fact islands of stars located outside the Milky Way. Since then, they've been piecing together observational data to try to understand how galaxies came to be, how they've evolved and what they can tell us about the Universe, its past and its future.

Cosmologists – scientists who investigate the nature of the Universe – don't just study individual galaxies; they look at how galaxies are grouped together. They've found that galaxies belong to local groups, which in turn form clusters of galaxies that are themselves contained in superclusters extending along a network of lines and creating filaments of galaxies. Each filament is interconnected

with other filaments and between them there are 'empty' voids. This large-scale structure of the Universe is what cosmologists call the 'cosmic web'.

SO JUST HOW BIG ARE THESE FILAMENTS?

Well, let's use the Milky Way for scale. Although we don't know the exact size of our Galaxy, we estimate it to be around 100,000 light years

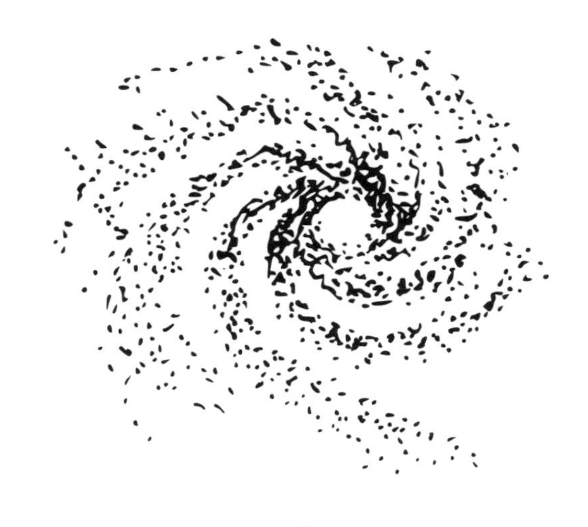

across, where a light year is the distance light travels in one Earth year – 9.5 trillion km (5.8 trillion miles). While the six-figure-light-year Milky Way is big compared to us, a cosmic filament can be millions, even billions of light years in length. In 2013, astronomers identified the largest structure yet discovered using gamma ray and x-ray telescopes. They called it the Hercules–Corona Borealis Great Wall and it spans over 10 billion light years!

WHY DO WE HAVE THESE STRUCTURES IN THE UNIVERSE?

For that we have to go back to the very beginning, shortly after the Big Bang, when the Universe was still a very hot and rather uniform soup of plasma. As

the Universe rapidly expanded it cooled down, allowing gravity to take over and begin causing matter to clump together. Over time, concentrations of matter attracted still more matter, creating regions with much more matter than others. Although the mechanism is not yet fully understood, astronomers believe it was the interaction between matter and dark matter that led to the formation of these denser regions in the Universe that eventually became the galaxies, groups, clusters, superclusters and filaments we see today. It appears that, overlaying the visible cosmic web, there's a much more massive dark cosmic web made up of the very mysterious dark matter.

PLAY WELL #2

Three custom Lego figures are on board the Juno *spacecraft currently orbiting the gas giant Jupiter. The figures represent the Roman gods Juno and Jupiter, and Galileo Galilei, the astronomer who discovered Jupiter's four largest moons (known as the Galilean moons) in the early seventeenth century.*

GALILEO GALILEI

INVENTIONS FROM SPACE

The enormous sums of money spent exploring outer space can sometimes seem difficult to justify; how does any of this directly affect us here on Earth? It turns out we have space research to thank for hundreds of pieces of technology that we now take for granted, ranging from the life-saving to the bizarre. Here's a look at some of the more unusual pieces of 'space-tech'.

MEMORY FOAM

Astronauts hurtling back towards Earth after a trip to outer space deserve a comfortable seat to lessen the impact. You know who else deserves comfort? The rest of us! Viscoelastic polyurethane foam was developed by NASA in 1966 for aircraft seats, but it wasn't long before its ability to mould to the shape of a human body was put to commercial use. Mattress manufacturers quickly adopted the material, giving all of us the chance to sleep in the

same level of peace and comfort as astronauts experience during re-entry...

GOLF BALLS

By the 1990s, dimples on the outside of a golf ball were a common feature. We knew they helped create a thin layer of air on the surface of the ball, allowing it to fly more smoothly through the air, but questions remained about the pattern and size of dimple that would work best. Using technology originally developed

FORE!

On 6 February 1971, Apollo 14 Commander Alan Shepard became the first human to play golf on the Moon. Using a makeshift 6-iron, Shepard made history by hitting two golf balls on the lunar surface, exclaiming that his second shot went for 'miles and miles and miles'. Remastered images have refuted this claim, as they show that the second shot only travelled 37m (121 feet) – still, it's not bad for someone wearing a cumbersome spacesuit!

to improve the aerodynamics of the Space Shuttle, the Wilson Sporting Goods company designed an aerodynamically optimised ball using a complex pattern of three differently sized dimples.

So, there you have it, scientific proof that the problem is your swing, not the ball...

MEAT ALTERNATIVES

Sending things to space is very expensive. In addition to all the hardware required for research, a significant amount of a mission's payload is dedicated to food to keep the astronauts alive. The solution seemed simple – design a way to produce food in space.

In a 1967 report entitled 'The closed life-support system', NASA scientists described a process combining waste carbon dioxide with a type of bacteria called a hydrogenotroph. These bacteria consume the carbon dioxide and produce a protein-rich liquid; essentially producing food from thin air. In a world where there is growing demand for meat alternatives to reduce the impact of emissions from livestock farming, interest in the idea has been

renewed. Commercial uses are in development and it may not be long before you're eating a space-age burger or sausage roll!

ON AN ASTRONOMICAL SCALE

In astronomy, our sense of size is often biased towards the first example we discover, or by trying to relate something to Earth or another intuitive scale. Often, we try to differentiate within broad categories by designating an object a 'giant' version of something when it is significantly larger than some standard, and 'dwarf' when it is smaller. Interestingly, these objects can sometimes have quite different properties as a result of their size, casting doubt on the wisdom of the original categorisation! Astronomy is confusing.

DWARF GALAXIES

Not all galaxies are vast cities of stars – some are a bit more modest. Dwarf galaxies are collections of stars that typically number in the billions, but some examples have a mere 1,000 stars within them. When compared to their larger cousins, which contain hundreds of billions or even trillions of stars, they may not seem like much. But the origins of these tiny

celestial neighbourhoods, which were likely some of the first galaxies to form during the cosmic dark ages or shreds torn off their larger parents, provide astronomers valuable insight into the evolution of galaxies.

WHITE DWARFS

As stellar deaths go, our own Sun's final moments will be rather sedate – no grand explosion, no cosmic burst of energy, just a series of sighs of gas as the star sheds its outer

layers, leaving behind its core. Small, hot and dense, this tiny Earth-sized remnant is known as a white dwarf. A single cup of the superheated material it is made from would weigh around 100 tonnes.

DWARF PLANETS

Poor Pluto. After 70 years as a beloved planet, it was demoted. And it wasn't even its fault. Dwarf planet is just the astronomer's name for something that fails to meet the requirements of being a planet. To make it into the big leagues, an object must be orbiting a star, have gravity strong enough to mould itself into a ball shape and be by far the largest thing within its orbital path. Having moons or asteroids nearby is fine, as long as you are clear of neighbours that can compete with you in size. Unfortunately, it's on this last point that Pluto and the other

dwarf planets fall down. Now, as definitions go, it's not exactly... exact. But for now, as far as astronomers are concerned, it means there are five identified dwarf planets in our Solar System (Pluto, Eris, Haumea, Makemake and the asteroid Ceres) with potentially dozens, or even hundreds, more out there. So at least Pluto's not alone...

RED DWARFS

The stars we can see in the night sky with our own eyes inhabit a remarkably small portion of our Galaxy. They represent some, but certainly not all, of the closest stars. The most common type is notably absent from our view: the red dwarf. Such objects only just fit the definition of a star at all, many are barely large and hot enough

to fuse hydrogen in their cores, which all true stars must do. But they make up around 70 per cent of the stars in the Galaxy and even the closest star to our own Solar System, Proxima Centauri, is one of them. So why can't we see them? Simple, they are very faint – at least as far as vast superheated balls of gas go. Even a few light years away they fade beyond our eyes' ability to see them.

PROXIMA CENTAURI

Proxima Centauri, discovered in 1915, is a small red dwarf star, with the distinction of being the closest star to the Sun. It is only approximately 4.2 light years away, and we have discovered three exoplanets in orbit around it (one is disputed at the moment). Sadly, none of these would be suitable for humans to live on, so we'll have to look further into space if we want to live somewhere beyond the Solar System.

BROWN DWARFS

In the awkward fuzzy area on the scale between large gas giant planets and small stars are brown dwarfs. Around the size of Jupiter, though a fair bit heavier, they never quite manage to heat up enough in their cores to fuse hydrogen. Many of these objects barely emit light that would be visible to our eyes and are instead most easily seen by telescopes sensitive to infrared. Recent studies have shown that many of our nearest galactic neighbours are these elusive stellar wannabes.

YELLOW DWARFS

Yellow dwarfs are medium-sized, medium-hot stars with moderate lifespans – not the most fascinating type of star from our point of view perhaps, but a very important one nonetheless. In fact, our own Sun is a yellow dwarf star. Relatively quiet, with few large flares erupting from their surfaces, stars like these tend to be the focus

of searches for life outside our Solar System. Despite the name, many yellow dwarfs are actually white, including our Sun. The yellow colour we associate with our luminous companion is

HABITABLE ZONE

Habitable zones are the regions around stars in which life could potentially exist. There are lots of variables used to calculate these and plenty of disagreement about the correct approach, but generally they represent regions where conditions allow water to remain in a liquid state, neither freezing into ice, nor boiling away as gas. A vital component in Earth-based lifeforms, liquid water may not be some grand universal ingredient for life, but it is probably the most sensible starting point in our search.

a result of Earth's atmosphere, which scatters blue light, making the sky appear blue and leaving the warmer colours in the light from the Sun untouched.

BLACK DWARFS

A solitary white dwarf is a fairly boring thing. With no fusion in its core and no nearby star to supply it with new gas, it has no mechanism to release more energy. Instead, its dim glow is sustained like an ember left over after a fire, slowly cooling down. But unlike an ember, which lasts but a moment, the energy of a white dwarf takes rather longer to radiate away. As it cools its colour will change, first to blue-white, then yellow, orange and, finally, to red, before it fades completely to a black dwarf. This process is expected to take a few quadrillion years or more. Given that our Universe is only some 13.8 billion years old, no white dwarf has yet had time to cool

down in this way. While black dwarfs may be effectively inevitable, until one actually occurs they remain theoretical in nature.

NAMING LUNAR FEATURES

The Moon is covered in craters of all ages and sizes. The naming theme for our Moon's craters is scientists, engineers and explorers. Some of the famous individuals lucky enough to have a crater named after them include H.G. Wells, the writer behind *The War of the Worlds* (1897), Caroline Herschel, an astronomer who discovered multiple comets and in the 1780s became the first documented woman in science to receive a salary for her work, and Jules Verne, the author of the 1872 novel *Around the World in Eighty Days*. There are others

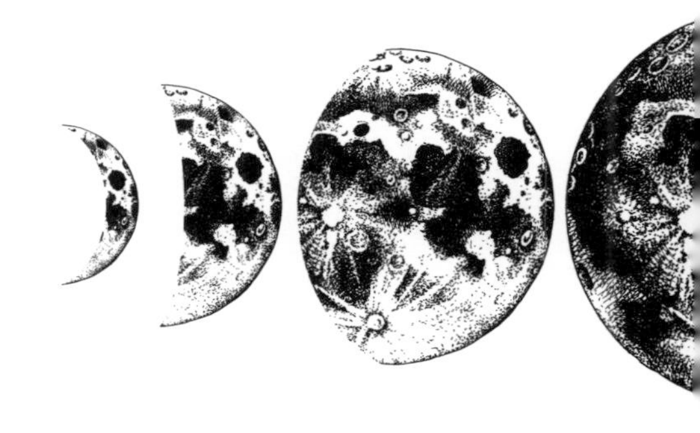

named for J. Robert Oppenheimer, the physicist who was part of the team that developed the atomic bomb in the United States in the 1940s, the tenth-century astronomer Muhammed Ben Geber Al-Battānī and Edwin Hubble (so he gets a space telescope and a crater!).

Maria (lunar seas) are darker regions on the Moon's surface, some of which are visible to the naked eye from here on Earth. Named after Latin words for weather and abstract concepts, they include Mare Tranquillitatis

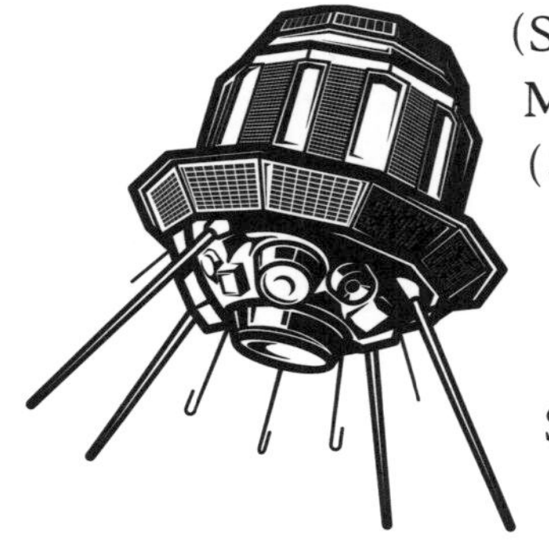

(Sea of Tranquility), Mare Humorum (Sea of Moisture), Mare Nubium (Sea of Clouds), Mare Serenitatis (Sea of Serenity) and Mare Moscoviense (Sea of Moscow).

One of these is not like the others. The Sea of Moscow is on the far side of the Moon, so it wasn't discovered until the Soviet spacecraft *Luna 3* first imaged the region in 1959. In celebration of their achievement, the team behind the mission suggested the name Moscow. This caused some consternation as it didn't fit the theme, but the concern was soothed by the suggestion in a 1961 IAU meeting that Moscow isn't just a city but also a 'state of mind'.

GALAXY CANNIBALISM

While Gaia Sausage sounds like it might be part of the Full English at your local gastropub, it's in fact what astronomers call the remains of a galaxy that was cannibalised by our own Milky Way over eight billion years ago. As with any good sausage, the fewer the ingredients the better – and Gaia's recipe is simple: stars, dust and gas. Its ingredients mixed with those that made up our Galaxy but weren't completely blended in, leaving a trace for modern astronomers to find. This has allowed them to confirm an astounding fact about our own Galaxy: it has evolved by merging with smaller galaxies.

When we study the Milky Way, we find that there are spherical groupings of tightly packed stars that astronomers call globular clusters, which

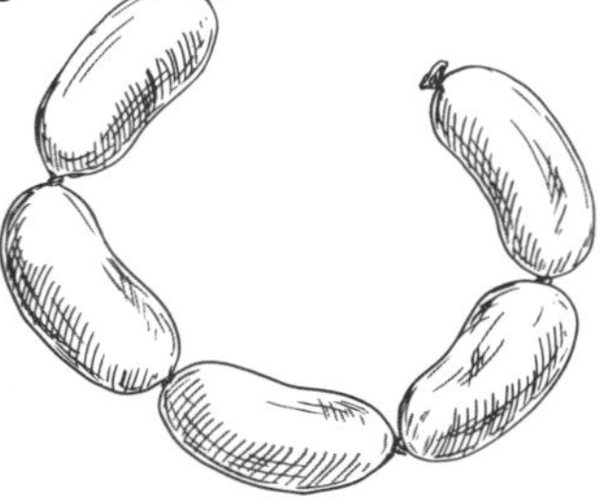

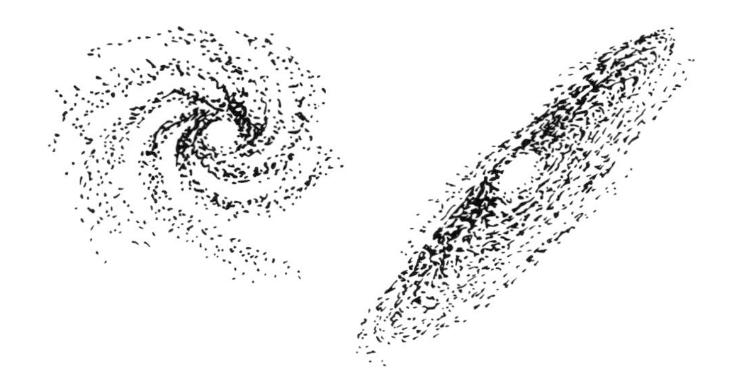

live in the halo of the Milky Way but not in its spiral arms, where our Sun lives.

WHAT DOES THIS HAVE TO DO WITH GALAXY CANNIBALISM?

By looking at the different globular clusters within our Galaxy, and studying their age, mass, composition and how they move, astronomers have established that their origins are different: not only are the clusters older than stars in the Galaxy's spiral arm, but some used to be the

ingredients of other sausages that were gobbled up by our own.

Galaxies come in very diverse shapes and sizes. Some are wispy, others look like clouds of stars and still others are interacting with their neighbours. Over the last 100 years or so, astronomers have put together all this information and concluded that galaxies grow and evolve through mergers with other galaxies.

ARE THESE GALAXY MERGERS A THING OF THE PAST?

Not quite. There are fewer mergers being observed in the present era, but the Milky Way is currently eating away at two dwarf galaxies, the Large and Small Magellanic Clouds. It is also moving towards our largest neighbouring galaxy, the Andromeda Galaxy, which is likewise moving towards us. The merger of

the Milky Way and Andromeda will be fully underway about 4 billion years from now. The merger process is not a violent one – actual collisions would be unlikely given the vastness of galaxies and the huge amount of space between their objects. Instead, it is the gravitational interactions that will reshape the galaxies and ignite star formation. The whole merger will last billions of years and will even include a union of the supermassive black holes at each galaxy's heart. At the end of the process, the Milky Way and Andromeda will have become one very large galaxy that astronomers are calling Milkdromeda. Sounds more like a chocolate bar than a sausage!

THE SOUNDS OF EARTH

NASA's twin spacecraft, Voyager 1 *and* 2*, are well on their way into interstellar space and both carry a unique time capsule intended to tell the story of Earth to any extraterrestrials who may come across them. The Voyager Golden Record, a gold-plated copper disc, consists of greetings in more than 50 languages, sounds of Earth, 115 analogue-encoded photographs and 90 minutes of music.*

MAGNIFICENT MOONS

We know of almost 300 moons in the Solar System at the moment and there may be yet more currently undetected. Here are five of the most interesting...

5. STRANGELY EARTH-LIKE TITAN

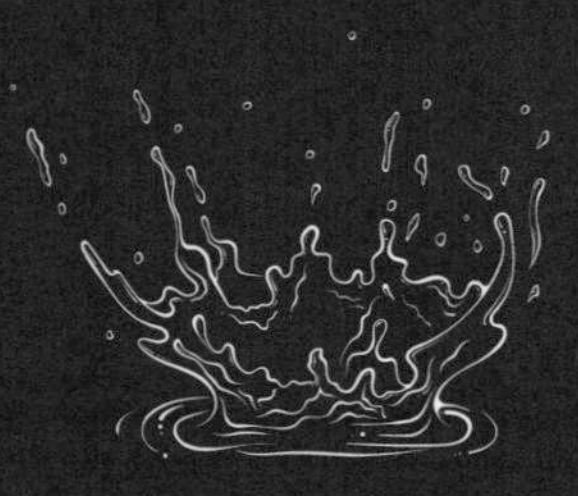

From a distance, Saturn's moon Titan just looks like an orange blob. The orange fuzziness is actually a thick, nitrogen-rich atmosphere – the only other place we've found one of those in the Solar System is here on Earth. And, perhaps even more excitingly, hiding beneath the clouds are stable lakes and seas, another feature uniquely shared by this moon and our planet.

But despite the similarities, both these features are also quite different from those on our home. Titan's air doesn't contain any oxygen, and its lakes aren't made of water. Instead, there are substances like liquid methane and ethane.

So close, and yet, so far!

4. TRITON – AN EX-DWARF PLANET?

The largest moon of Neptune, Triton is covered in cantaloupe terrain (that looks like the skin of a melon) and cryovolcanoes that spew out ice instead of lava.

It is the moon's orbit, however, that gives one of the biggest clues to its origins. Triton orbits Neptune in the opposite direction to the planet's spin, unusual for such a large moon. Combined with the apparent disruption it has caused to other Neptunian moons, it seems that, rather than forming alongside the planet itself, Triton was probably captured from elsewhere – most likely the Kuiper belt, where most of the dwarf planets are found.

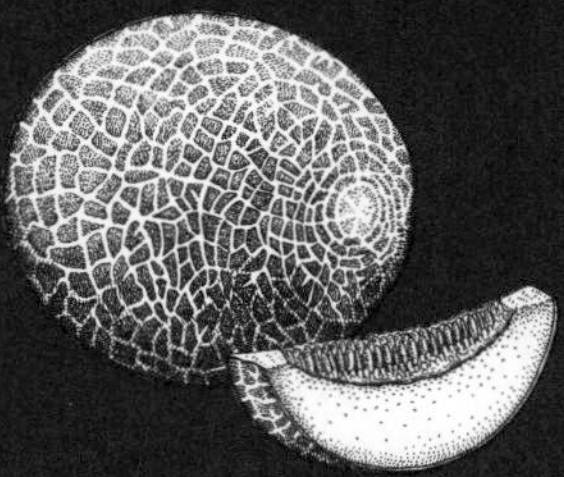

3. PLUTO AND CHARON'S COSMIC DANCE

With about half the diameter of Pluto and roughly one eighth its mass, Charon has the closest parent-to-moon size ratio of any planet or dwarf planet satellite in the Solar System. Being so relatively large, Charon pulls on Pluto enough to make the dwarf planet do a teeny orbit of its own around their combined centre of gravity. Because of this, they can almost be considered a binary system (in which two objects orbit each other).

The two are also tidally locked, meaning that Charon always has the same side facing towards Pluto and vice versa.

Is this a romantic twirling waltz? Or two enemies hiding their backs from each other?

2. FRANKEN-MOON MIRANDA

Uranus's moon Miranda has a jigsaw-like surface that looks as though someone smashed it up before trying to put it back together again. This may be exactly what happened – debris may have crashed into the moon, meaning that Miranda had to use the gravity of the pieces to pull itself back together. Alternatively, maybe subsurface ice melted and refroze, moving the rocky surface around.

Either way, Miranda's past has left it with some pretty spectacular features, including the tallest cliff in the Solar System. It's so high that it would take ten minutes for a stone dropped from the top to reach the bottom!

1. PAN THE CELESTIAL RAVIOLI

Pan lives inside the rings of Saturn. With a diameter of 28km (17 miles), it's one of the biggest objects found there, surrounded by hundreds of icy chunks of ring material. It has carved out its own path inside the rings, using its gravity to move material aside, and inducting Pan into a group known as 'shepherd moons'. But all this ring-sculpting has reshaped Pan too. Some of the ring material it swept up has become stuck around the moon's equator, creating a large ridge that looks a bit like a tutu. This distinctive shape means Pan is often compared to similarly shaped food, everything from walnuts to ravioli and empanadas!

HONOURABLE MENTIONS

- Phobos and Deimos, Mars's moons, are often called 'space potatoes' because they look like, well, potatoes!
- Orbiting Saturn, the planet most famous for its rings, is Rhea, a moon that possibly has rings of its own.
- And then there's Hyperion, the first non-round moon ever discovered. No one knows for sure why it looks like a sponge.

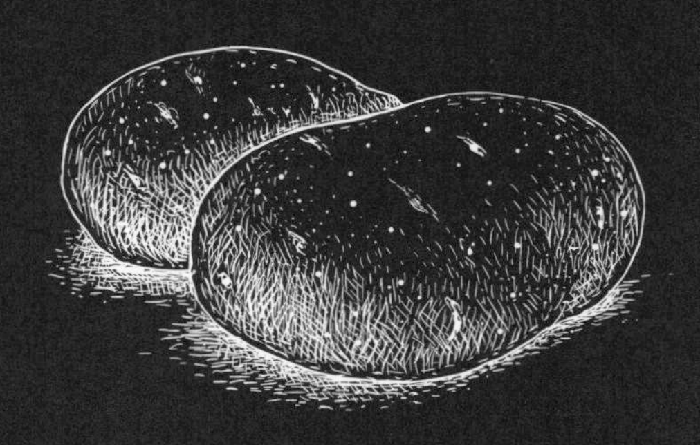

WHEN THE SKY FALLS TO EARTH

Over its near 5-billion-year history, Earth has been subjected to countless impacts from space. In fact, even nowadays, an estimated 44 tonnes of space rocks – termed meteoroids – collide with Earth every day. The overwhelming majority are tiny, most no larger than a pebble. But as our Jurassic ancestors can attest, the largest of these rocks can have much more severe consequences...

Let's look at some of the biggest impact events ever to occur on Earth:

2013 CHELYABINSK METEOR

The term 'meteor' is used to describe a meteoroid once it has entered Earth's atmosphere. In 2013, a 20m (66ft) wide meteor arrived over the town of Chelyabinsk in Russia. Hurtling undetected through the air at incredible speed, the immense pressure build-up caused it to explode before it hit the ground. The explosion produced roughly 30

times more energy than an atomic bomb, injuring thousands but miraculously killing no one.

Perhaps the unique thing about this event is that it's the only one on our list you can find footage of online!

1908 TUNGUSKA EVENT

The biggest impact event in recorded human history happened in 1908, close to the Podkamennaya Tunguska River in northern Siberia. At just after seven o'clock in the

morning, eyewitnesses heard a thunderous crash and saw a flash like 'a second sun'. In an instant, everything within 2,150 square km (830 square miles) – an area as large as the country of Monaco – was flattened or set alight. There were even reports of a shockwave powerful enough to smash windows hundreds of miles away.

When scientists finally arrived at the remote area in 1927, they were confused by the absence of an impact crater at the site. Very soon, a myriad of weird and wonderful theories filled this void, with everything from extraterrestrials to mini black holes suggested as explanations. Thanks in part to observations of the Chelyabinsk meteor over 100 years later, we now have a much clearer

understanding of these air-burst impact events.

CHICXULUB

To go bigger – and we can go way bigger – we must turn our attention to pre-history. Many clues we have about impact events in this period are gleaned from the permanent scars they left on Earth's surface, scars that we can see today as craters.

The Chicxulub crater lies just off the coast of the Yucatán peninsula, in modern-day Mexico. At about 180km (112 miles) across, it is famous for being ground-zero of the Cretaceous–Paleogene extinction event or, put simply, the extinction of the dinosaurs.

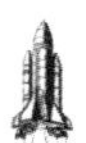

ASHES TO ASHES #2

In 2012, the cremated remains of Canadian actor James Doohan (1920–2005), famous for his portrayal of the affable chief engineer Montgomery 'Scotty' Scott in the original series of Star Trek, *were launched into space. This was not the first time his ashes had been sent to space. Laminated photographs containing his remains were reportedly smuggled onto the International Space Station in 2008 by Richard Garriott, the sixth private citizen to fly into space.*

The space rock in question is estimated to have been an asteroid roughly 10km (6 miles) in diameter. According to NASA, impacts like this happen every 250 to 500 million years. At only 66 million years old, the Chicxulub impact crater is hopefully a sign that we are safe from giant flaming extinction balls for the time being!

VREDEFORT

The Vredefort impact basin, a 300km (186 mile) wide crater in modern-day South Africa, is the largest verified impact to have occurred on Earth. It is thought to have been formed over 2 billion years ago by a space rock nearly 25km (16 miles) in diameter.

It isn't clear exactly where this space rock originated, but from geological studies of the area we know that enough energy was released to turn parts of Earth's crust to liquid. Thank goodness us humans weren't around at the time!

And there you have it, a whistle-stop tour of our planet's tumultuous relationship with cosmic projectiles. If you feel a little more concerned about the fragility of our species than you were before, remember that in human timescales these kinds of events are astronomically rare, so you can rest easy knowing the sky isn't going to fall in for the next few million years at least.

DISORDER AND CHAOS

The discovery of Eris in 2006 kicked off a debate among astronomers, as more planetary bodies of similar size and with similar orbits were being discovered in Pluto's region of the Solar System. Rather than label them all planets, a new category was invented, and both Eris and Pluto were shunted into the dwarf planet club. The move made sense from an organisational perspective but upset a surprising number of people.

In honour of the chaos caused by the finding, the team suggested the object be given the name Eris after a Greek goddess of discord and strife. Eris's moon, Dysnomia, is named after a daughter of Eris who represents lawlessness.

MAY THE FORCE BE WITH YOU

A lightsaber prop used in the Return of the Jedi *film was taken into space on board the space shuttle* Discovery *in 2007 to celebrate the 30th anniversary of the release of the original* Star Wars *trilogy.*

Continuing the god and goddess naming theme, other dwarf planets include Ceres (after the Roman goddess of corn and the harvest), Pluto (after the Roman god of the underworld), Makemake (after the Polynesian creator god) and Haumea (after the goddess of fertility and childbirth in Hawai'i).

There are lots of god and goddess names not yet used by the IAU, but there are probably also plenty of dwarf planets still to be discovered and named in the future.

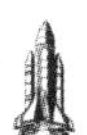

WE USED TO THINK

... THAT EARTH WAS THE CENTRE OF THE UNIVERSE

Throughout history and across cultures, a variety of geocentric models (that is, placing Earth at the centre) of the Universe have been favoured, but efforts to correct what we now know to be an unfortunate mistake are more varied than many realise. The earliest recorded examples of a model of the Universe that didn't place us at the centre date to a few hundred years BCE and the ancient Greeks.

Pythagorean philosopher Philolaus (*c.*470–*c.*385 BCE) suggested that all things orbited some central fire. This may sound familiar, except he didn't mean the Sun, which he claimed also revolved around the centre, whatever it was. Only 100 or so years later, Aristarchus of Samos (*c.*310–*c.*230 BCE) proposed a system where the cosmos revolved around the Sun, which he believed to be several times larger than Earth.

This heliocentric (centred on the Sun) view of the Solar System was revisited a number of times over the next two millennia but wasn't widely accepted until the Copernican Revolution in the sixteenth century CE. Between them, Nicolaus Copernicus, Johannes Kepler and Galileo Galilei provided the model and observations that forced humanity to reevaluate its place in the Universe. As time

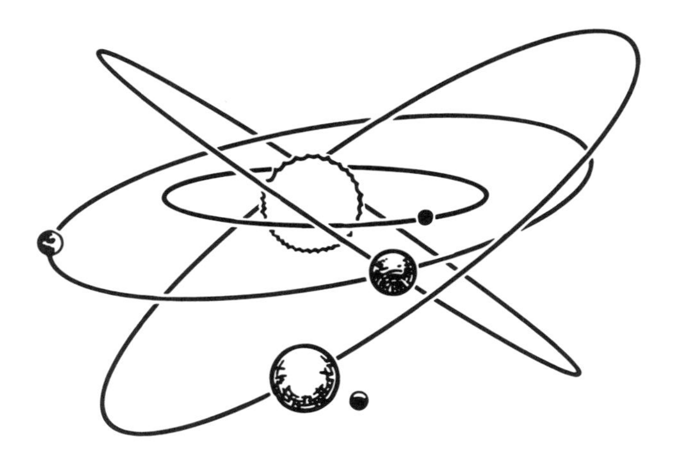

went on, a deluge of observations made with rapidly improving telescopes showed that not only were we not in the centre of the Solar System, we weren't even in the centre of our Galaxy. Nor was the Galaxy alone, being one of a great many. Further, the Milky Way wasn't particularly special, having no exceptional place in the Universe as far as we can tell.

From the centre of all creation to an insignificant speck in a Universe filled with insignificant specks, the human ego has taken quite a beating in the past few hundred years.

...THAT GRAVITY WAS 'SIMPLE'

It would not have taken early humans many throws of a rock to realise that it never just stayed up in the air, but knowing what caused it to fall again was a much more complicated matter.

Little by little, scientists chipped away at the problem, though the stories of how they did so are often exaggerated. Galileo definitely didn't drop cannonballs off the Leaning Tower of Pisa to prove that heavy objects don't fall faster than light ones, and Newton probably didn't get struck on the head by an apple while working out how the Moon orbits Earth.

By the early 1800s, we thought we had the Law of Gravitation sussed, as was demonstrated when astronomers and mathematicians like Le Verrier successfully used it to predict where to find Neptune. But tiny inconsistencies in the orbits of the planets were clues that there was more to it and when Albert Einstein published

his General Theory of Relativity he claimed that gravity was the result of the fabric of the Universe being distorted and stretched.

To say it's complicated is like saying the Pacific Ocean is a bit damp. From gravitational redshift to time dilation, Relativity introduced a whole barrage of new concepts that have mostly been proven correct.

Now more than a century's worth of physics undergraduates have longed for the simpler days, when the world had never heard of the Riemann curvature tensor or the Friedmann–Lemaître–Robertson–Walker metric...

...THAT THE UNIVERSE WAS STATIC AND UNCHANGING

Well, some people did anyway.

The concept of a changing and evolving Universe is hardly new, but some people have worked under the assumption that the Universe has always been around and always will be, a so-called 'Steady State'.

In the 1920s the Universe was shown to be expanding, which was a hard pill to swallow for those who believed the Universe had always existed. An expanding Universe implied that galaxies were closer together in the past and suggested a possible beginning – the Big Bang.

That's not to say advocates of the Steady State gave up. Some adopted something called the 'perfect cosmological principle'. The cosmological principle is an unproven but generally accepted assumption about the state of the Universe. In simple terms it means that no place, nor direction, in the

NEW BREW FOR THE CREW

The world's first zero-gravity espresso machine, nicknamed ISSpresso, debuted on the International Space Station in 2015. Developed in a joint venture between the Italian Space Agency, engineering firm Argotec and the Italian coffee company Lavazza, the experimental machine provided a much welcome alternative to the freeze-dried instant variety typically on offer. Astronaut Samantha Cristoforetti was the first to fire it up and enjoy space-brewed coffee.

Universe is fundamentally different to any other, and if you look at a large enough slice of the Universe it will look effectively the same as any other slice.

The perfect cosmological principle adds one extra bit, the more controversial one: that the Universe also looks the same at all times. To stop the expansion of the Universe betraying that third principle, proponents suggested a model in which matter was continuously created and expelled from some central point in the Universe, backfilling the void left empty by receding galaxies so that, at any point in time, the Universe would appear to all intents and purposes the same.

Unfortunately, despite valiant attempts to make it work, every Steady State model has

failed to account for all our observations of the Universe and thus, for now at least, the Big Bang model remains the best description we have for the evolution of our Universe.

... THAT STARS WERE COLLAPSING UPON THEMSELVES

In the 1800s, the question of how stars were powered was a source of considerable debate. At the time, the concept of energy conservation – the idea that energy cannot be created or destroyed, only changed from one form into another – was relatively new. If energy could not be created, then where was the Sun (being the star of immediate concern) getting its power from?

The problem was that not many sources of energy were known. There was chemical energy, like burning fuels, but that could only power something as luminous as the Sun for a few thousand years. Gravitational energy was another possibility – dropping something like a meteorite onto a planet would release a lot of energy. But that would only last a million years or so, and at that time the Sun was estimated to be many millions of years old.

In the mid-1800s, physicists William Thomson (known as Lord Kelvin) and Hermann von Helmholtz suggested another possibility. What if the Sun was constantly collapsing upon itself, slowly shrinking as it deflated like a balloon? This would gradually release energy, keeping the Sun hot and bright.

Based on estimates of the Sun's size and mass, it was estimated that this could power

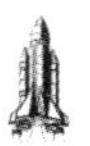

the Sun for up to 20 million years! The only problem was that geologists had recently shown that Earth was about 2 billion years old, possibly more...

Kelvin and Helmholtz claimed the geologists were wrong. Unfortunately for the pair, they weren't, and in the 1920s astronomers concluded that a newly discovered source of energy, nuclear fusion – the process of combining light elements together to make heavier ones – was the source of the Sun's radiance.

SPACE SIGNALS

As humans we have a deep-rooted desire to communicate, to connect with one another and make ourselves heard. As technology has developed, we have naturally extended this desire into the cosmos in the hopes of establishing communication with extraterrestrial civilisations – if they should wish to talk to us, that is.

The most efficient way we've found to attempt this is via radio signals, and for good reason. For a start, they travel at the speed of light (300,000km/s, or 671 million mph), making them mind-bendingly fast. As well as this, the long wavelengths of a radio signal can traverse the dust clouds of space with minimal disruption, allowing them to continue on and on forever or until

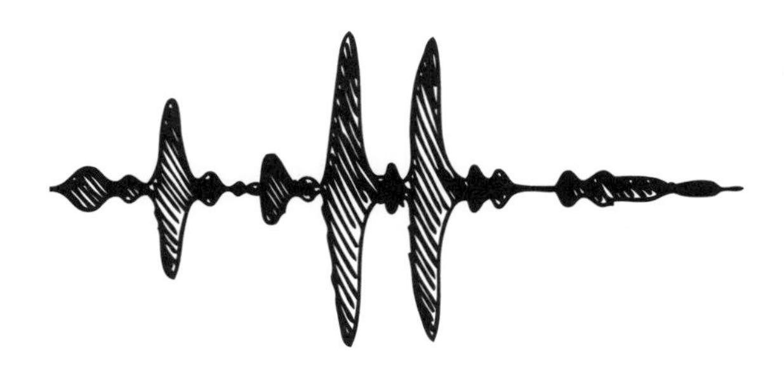

they reach their eventual destination, perhaps billions of light years away.

While there haven't been any replies yet, here are a few of the notable radio signals travelling through the Universe at this very moment:

THE MORSE MESSAGE

In 1962, humankind sent its first ever radio message intended to reach extraterrestrials. It was transmitted by the Yevpatoria Planetary Radar in Crimea – its destination, Venus. Just

three words were broadcast, all in Morse code: 'mir' (Russian for both 'peace' and 'world'), Lenin and USSR. Although there was no reply, the signal did bounce off the surface of the planet and is currently heading towards the star HD 131336 in the constellation Libra. Perhaps the message will have better luck there.

THE ARECIBO MESSAGE

One of the most famous examples of a human-made space signal is the Arecibo Message, which was beamed from the Arecibo Telescope in Puerto Rico in 1974. It was sent towards Messier 13 (NGC 6205), a group of tightly packed stars often known as the Hercules Globular Cluster. The contents of the coded message was designed by astronomy heavy-hitters including Frank Drake and Carl Sagan, and included the numbers one to ten, the atomic numbers of the elements that make up human DNA, a graphic of the planets of the Solar System (including the now dwarf planet Pluto) and much more.

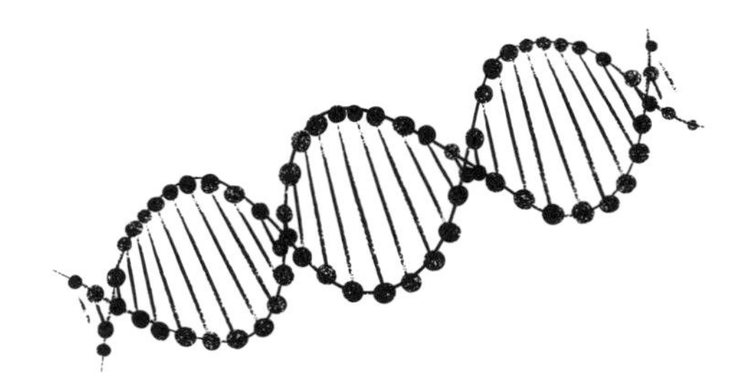

The project was devised more to demonstrate human technological advancement than in the hope anyone would receive the message, which makes sense considering it won't reach its destination for at least another 25,000 years. Perhaps an update message regarding Pluto is in order, though.

ACROSS THE UNIVERSE

Some radio signals are sent into space purely for fun. To celebrate the 40th anniversary of the recording of the Beatles song 'Across the Universe' in 2008, it was beamed beyond Earth by NASA. It became only the second piece of

music to be transmitted deliberately into deep space; the first was a theremin concert made for a series of transmissions called the Teen Age Message in 2001. The Beatles track is currently on its way to Polaris, a star roughly 433 light years from Earth, meaning it will take the signal 433 years to get there. With a 'long and winding road' ahead, all we can do now is 'let it be'.

MISSION TRANSMISSION

Messages and songs are one thing to send into space, but how about an entire radio programme? In 2022, British radio station Fun Kids earned a Guinness World Record for having the first radio programme beamed into deep space (on purpose). The record-breaking show, titled *Mission Transmission*, featured

the voices of hundreds of children with messages to potential alien listeners, as well as some of our Royal Observatory astronomers. The launch was celebrated at Royal Observatory Greenwich with the help of British astronaut Tim Peake. It will be quite a few years before the signal reaches another star system, but perhaps in the far future on a distant planet an alien radio engineer will tune in and break a world record of their own by doing so.

NAMING EXOPLANETS

There are more than 5,000 confirmed exoplanets in our Galaxy, so it's really going to stretch our collective creative-thinking skills to come up with names for them.

Currently, most exoplanets are simply given a catalogue number linked to the telescope that discovered them. For example, TRAPPIST-1 d was found by the Transiting Planets and Planetesimals Small Telescope (TRAPPIST). The star was named TRAPPIST-1 and the third planet discovered orbiting it is 1 d (the system starts with b, then c, and so on).

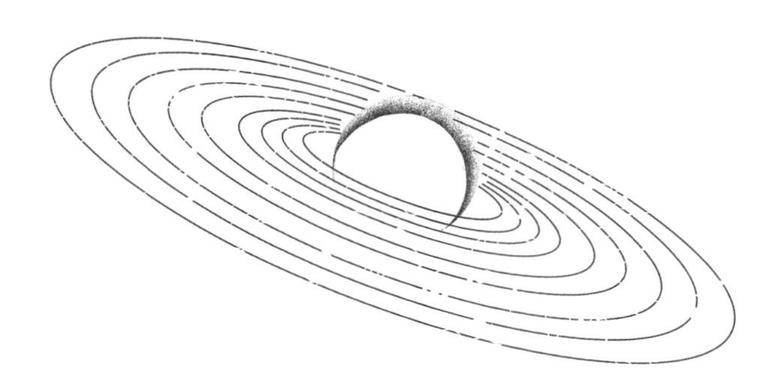

Some exoplanets do have proper names ratified by the IAU, which has held multiple competitions where the public can put forward names. There are still a few requirements: you can't name a stellar system after a living person, after your pet, or after anything principally commercial in nature (so no Coca-Cola world).

Named exoplanets include:

- Poltergeist – one of three planets (the others are Phobetor and Draugr) orbiting a neutron star named Lich. Planets orbiting neutron stars aren't thought to be particularly habitable places, possibly inspiring the ghoulish names.
- Perwana – meaning moth in Urdu, referencing the fact that just as moths circle lights, planets orbit stars.
- Abol – the first of three rounds of coffee in the traditional Ethiopian coffee ceremony.

- Enaiposha – a word from the Maa community in Kenya that means 'an expression of awe at the tumultuous nature of a large amount of water'. Enaiposha is a Neptune-like exoplanet, but the volume of water there is currently unconfirmed.
- Victoriapeak – the name of the peak overlooking Victoria Harbour in Hong Kong.
- Dimidium – Latin for half, a reference to the fact that this planet's mass is half that of Jupiter.
- Finlay – after Carlos Juan Finlay (1833–1915), a Cuban doctor who pioneered research into yellow fever. This also means that if you know anyone named Finlay you can now immediately inform them that there's a planet named after them!

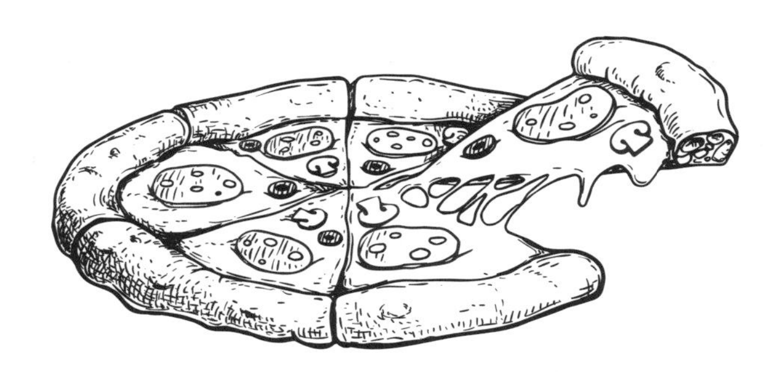

SPECIAL DELIVERY

Back in 2001, Pizza Hut became the first company to deliver a pizza to the International Space Station. The smaller-than-standard, custom-made salami pizza and the delivery fee cost the company over $1 million (about £696,000) at the time.

SURPRISINGLY USEFUL OBJECTS

Getting humans into space is an incredible scientific and technological achievement, a massive engineering challenge that involves thousands of experts from all around the globe.

This isn't about the marvels of design or the technological brilliance of the Apollo spacecraft, however, but some small and surprisingly useful objects that played their part in getting the astronauts safely home again.

A PEN

In July 1969 Neil Armstrong and Buzz Aldrin became the first humans to land on the Moon, spending just under 24 hours on the surface. They then had to become the first humans to safely take off from the Moon again. Unfortunately, at some point when manoeuvring in their bulky spacesuits and life support backpacks in the lunar module, one of the astronauts accidentally snapped off the

switch marked 'engine arm'. This was effectively the switch to turn on the engine, needed to lift off. This would have been a serious problem if not for a trusty felt-tip pen – Aldrin pushed this in and wiggled it around, managed to arm the switch, power the engine and successfully and safely take off again.

DUCT TAPE AND BOOK COVERS (AND A CANNIBALISED SPACESUIT)

You may have heard of the problems faced by the crew of Apollo 13, the mission in which the famous, but often misquoted, words 'Houston, we've had a problem' were spoken. 56 hours after lift-off, an explosion damaged

PLAY WELL #3

Four Lego astronaut minifigures hitched a ride to the Moon in 2022 on board the **Orion** ***spacecraft as part of the Artemis I mission. They were called Kate, Kyle, Julia and Sebastian.***

the spacecraft and made continuing the mission and landing on the Moon impossible. To survive, Jim Lovell, Jack Swigert and Fred Haise moved to the lunar module (the part of the spacecraft designed to take two astronauts down to the Moon's surface) and used it as a lifeboat for the return journey to Earth. However, the systems meant for removing carbon dioxide from the air weren't designed to cope with three people for four days and the amount of carbon dioxide in the air became a dangerous problem.

Using materials scavenged from the service module, a spacesuit valve and the cardboard

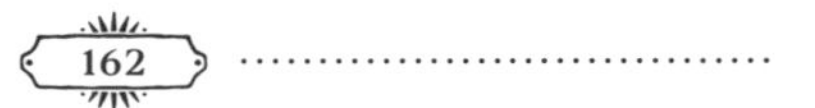

covers ripped off their instrument manuals, the crew on the spacecraft, with help from the engineers on the ground, managed to create a life-support system fix which lasted the rest of the journey home, allowing the astronauts to splash down safely in the Pacific Ocean on 17 April 1970.

LAMINATED MAPS

The later Apollo missions (15, 16 and 17) used a lunar rover to explore more of the Moon's surface, travelling much further from their landing sites than previous missions. Apollo 17 astronauts Jack Schmitt and Gene Cernan hold the records for the furthest distance travelled from their landing site, almost 8km (5 miles), and the longest total drive, 36km (22 miles) over four hours.

The Moon's surface is covered in a layer of dust known as regolith. This material was a constant hazard, easily kicked up by the astronauts' activities. It became an even bigger problem when Cernan managed to snap off one of the rover's fenders. Without a fender,

MOON ROCKS

Lunar rocks give scientists an opportunity to learn about formation processes on the Moon itself and also to compare the material to that on Earth and gain insights into the formation of the Solar System (and by extension, other planetary systems). In an amazing example of forward-planning, some Apollo-era Moon rock samples were left in storage for 50 years. Astronomers knew our ability to analyse samples would improve and decided to rely on this for new insights in the absence of more Moon-landing missions. Scientists are working on these rocks right now...

dust was thrown everywhere, obscuring the astronauts' vision and covering important equipment.

Fortunately, the astronauts had a trusty roll of duct tape and they taped four maps together and clipped them to the rover. This fix lasted for the remainder of the surface activities, allowing the astronauts to collect 115kg (18 stone) of rocks, the most of any Apollo mission.

This was the last time humans made it to the Moon during the twentieth century. We hope that future lunar exploration will go smoothly, but perhaps it's wise to remember the surprisingly useful objects used in the past and maybe pack an extra roll of duct tape, just in case.

THE END OF EVERYTHING

All good things must come to an end. In our lifetimes, we think about things like the end of a holiday or saying goodbye to school friends, but these events are miniscule in the timescale of the Universe. Let's consider what will happen in billions and trillions of years' time and what the end of everything might look like...

THE END OF EARTH (~5 BILLION YEARS)

Our Sun is a huge ball of gas, and our source of heat and light. It is one reason why Earth is a habitable place for us, but it will also be the reason for the unfortunate fate of our planet. The Sun gives off energy produced by the nuclear fusion reactions in its core, where hydrogen is converted into helium. Currently, our Sun is around halfway through its hydrogen supply. Eventually, though, the hydrogen will run out and when it does the Sun's core will begin to collapse inwards. Its outer layers will then expand, forming a red giant star.

It is unknown how much the Sun will expand, but current estimates suggest that it will engulf Mercury, Venus and maybe even Earth. 'Maybe' is good, right? Wrong! Even if it doesn't reach Earth, the extreme heat from the then-enormous Sun will be enough to evaporate our planet's oceans, increase the surface

temperature to hundreds of degrees and end all life as we know it.

THE END OF THE SOLAR SYSTEM (~8 BILLION YEARS)

This is only the beginning of the end of our Sun's life. After its red giant phase, the star's outer layers will continue to expand into space, so much so that they will create a planetary nebula (nothing to do with planets!). The outer layers will leave behind a white dwarf, the old star's core.

Through this process, the Sun will lose about half its mass, causing the gas giant planets (Jupiter, Saturn, Uranus and Neptune) to migrate outwards into wider orbits. They will become less stable as they move further from the source of gravity keeping them in place, making it easier for another star to scoop them up.

Although space is big, over large timescales gravitational interactions with other star systems passing by can cause dynamic instabilities and potentially even eject planets from their systems entirely. These 'rogue' planets may journey for eons relatively undisturbed, but they could be captured by another star system, entering into orbit around a new star. The number of variables involved make this kind of change very difficult to predict even roughly, but stellar encounters may eventually leave the white dwarf remnant of our Sun all alone.

THE END OF THE MILKY WAY (~100 BILLION YEARS)

We know that several billion years from now the Milky Way will eventually merge with our closest large spiral galaxy, Andromeda, but it might not be the only merger our Galaxy goes

through. Due to gravity, the majority of galaxies travel in groups. The Milky Way is part of the Local Group, comprising approximately 40 galaxies travelling through the Universe. It is likely that in tens of billions of years, these galaxies will all merge to form one big elliptical

PLANETARY NEBULAE

Through early telescopes, astronomers saw fuzzy roundish shapes and promptly called them planetary nebulae. This has annoyed other astronomers ever since. In fact, the shapes are expanding layers of dust and gas pushed outwards by mid-mass stars during their red giant phase. A gentle end, as far as stars are concerned – they are essentially enormous explosions on any sort of human scale, destroying any planetary system that might exist around the star. From afar they look pretty, but they don't even look like planets!

galaxy, leaving no evidence of their separate pasts.

THE END OF COSMOLOGY (~150 BILLION (150×10^9) TO 10^{100} YEARS)

The Universe itself is expanding. We observe galaxies moving away from us and the more distant they get, the faster they move. At extremely large distances, galaxies may recede faster than the speed of light. If this continues, the Universe will be expanding so quickly that the light from distant galaxies will be unable to reach us any more and they will no longer be visible in the observable Universe. Eventually, all that remains will be black holes, neutron stars and some black dwarfs (the fate of stars like our Sun).

Matter is generally stable, but over very long periods it is believed even fundamental atomic particles – the most basic building blocks of

matter – may break down, leaving only black holes. Black holes have a lot of mass, so they will remain for a long time. Although uncertain, British theoretical physicist Stephen Hawking (1942–2018) predicted that even black holes will slowly give out radiation, losing mass until they eventually 'evaporate', indicating the beginning of the end.

THE END OF THE UNIVERSE (OR IS IT?) (~10^{200} YEARS)

With the last black hole gone, we are at the end of the Universe's timeline... maybe. We don't really know what the Universe truly looks or is like, so its end could manifest in many ways.

If the Universe continues to expand, energy will be spread out so much that eventually all areas within it will remain at the same temperature. We call this the Big Freeze, when interactions between particles will cease and the Universe

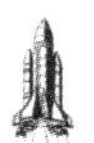

will remain at a frozen standstill forever. But if you don't fancy the idea of a cold, dark, barren Universe, then don't worry! There are still more morbid possibilities for how the Universe will end.

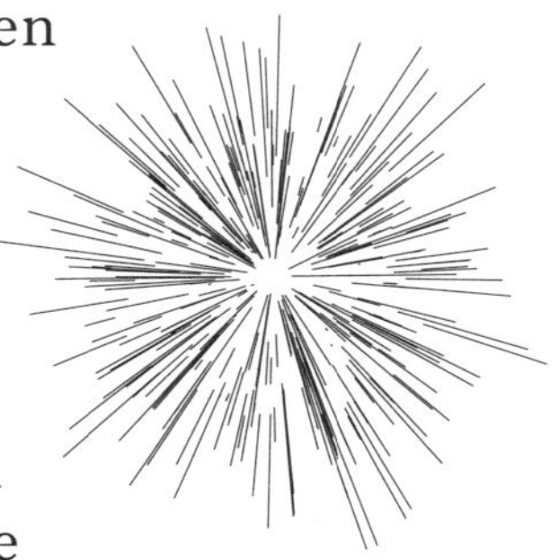

As we know, the Universe is accelerating in its expansion. If it carries on expanding at this ever-increasing rate, it might expand so fast that gravity can't hold anything together any more. The Big Rip will occur and everything from galaxies to atoms will be ripped apart. All that will be left will be single particles travelling through a seemingly empty, timeless space.

There might still be hope, as some scientists say that the Universe won't expand forever. The Big Bounce theory suggests that, in the future, the

amount of matter in the Universe will be so great that gravity will cause the expansion to slow down and stop. Then, it will reverse, causing the Universe to contract until it combines into a tiny point called a singularity before expanding out again, triggering another Big Bang and resulting in the birth of a new Universe. This process could happen again and again, our Universe and future universes stuck in a never-ending cycle of contraction and expansion.

It is worth pointing out that the timescales here leave plenty of time for fun, including studying more astronomy! Given everything people have learned since our earliest days (hardly the blink of an eye as far as the Universe is concerned), it would be very surprising if we didn't have a huge amount more to unearth. Perhaps the most exciting knowledge is still out there waiting to be discovered and will cast our whole understanding of the cosmos in a brand-new light. Maybe you'll be the one to transform what we think!

LIST OF CONTRIBUTORS

Since 1675, Royal Observatory Greenwich has been home to astronomers, and we love facts and stories about space! But the Universe is a big place, so the following people all contributed the bits they like best:

Affelia Wibisono

Anna Gammon-Ross

Ed Bloomer

Greg Brown

Jake Foster

Jess Lee

Jess Sells

Julienne Hisole

Luke Hand

Patricia Skelton

Tania de Sales Marques